YOUR KNOWLEDGE HAS VALUE

- We will publish your bachelor's and master's thesis, essays and papers

- Your own eBook and book - sold worldwide in all relevant shops

- Earn money with each sale

Upload your text at www.GRIN.com and publish for free

Imprint:

Copyright © 2016 GRIN Verlag, Open Publishing GmbH
Print and binding: Books on Demand GmbH, Norderstedt Germany
ISBN: 9783668281240

Mohamed Abdelwhab

New explanation of Michelson-Morley experiment

GRIN Publishing

New explanation of Michelson–Morley experiment

October 6 University, Egypt

Abstract

We present an alternative to Einstein's explanation for the null result of the Michelson–Morley experiment. We conceptualize a concept that states, "time's arrow is moving along a direction at a speed equal to the speed of light." As a result, we find "there is a gap between space and time"; thus, the position of a body at a given point does not necessarily imply that the impact of that body must initiate from this point. We use this idea to explain the reasons for quantum entanglement despite the constancy of the light's speed and the Michelson–Morley experiment.

Keywords

Michelson–Morley experiment, special relativity, Lorentz transformation, ether theory, spooky action at a distance, quantum mechanics

Content

1. Introduction

In 1887, Michelson and Morley performed the well-known Michelson–Morley experiment to determine the speed of the earth relative to that of the luminiferous ether [1, 2], which was considered the fundamental substratum of space and believed to be the medium of light propagation [3]. We recommend referring to a video prepared (https://www.youtube.com/watch?v=KTsD3vI04-g) to help simplify the understanding of our study and the concepts presented. The idea of the experiment can be summarized as follows: "The motion of the earth in the ether at velocity v generates an ether wind with the same velocity (scene 1); therefore, if we successfully measure the effect of the ether wind on the motion of light, it will serve as a strong evidence for the existence of ether (scene 2)."

The null result of the Michelson–Morley experiment is considered strong evidence against the ether theory [4] (scene 3) and is unexpected according to Galilean physics. In 1892, Lorentz first explained this null result in an attempt to conserve the ether theory; he suggested that the length of a body/object in the direction of motion contracts by an amount equal to γ (the Lorentz factor) because of a postulated similarity between molecular cohesion forces and electrostatic forces [5] (scene 4). The Lorentz transformations are a set of mathematical equations [6] used to correlate the space and time coordinates of a moving system to determine the space and time of another system when two observers (each in either system) are moving relative to each other. The Michelson–Morley experiment can be explained by these transformations; the length of an object along the direction of motion contracts (by a factor equal to γ) [7] while transforming to a moving frame. Consequently, the speed of light is identical in all frames, thus yielding the null result of the Michelson–Morley experiment.

Following this, in 1905, Einstein posited the non-existence of the absolute medium of ether and introduced the theory of "special relativity," which is based on two postulates: first, the laws of electrodynamics and optics are valid for all frames of reference, and second, the speed of light is constant regardless of the motion of the light source [8] (scene 5). Einstein deduced the Lorentz transformation from these two postulates. Consequently, he suggested that the length of moving bodies contracts along the direction of their motion and that the bodies undergo time dilation [9, 10] (scene 6); therefore, the result of the Michelson–Morley experiment is negative. Special relativity introduces a different system in which space and time are not absolute for all inertial frames [11]. Rather, they are relative to the frame of reference [12] unlike in the Newtonian world, where space and time are absolute for all inertial frames.

Despite the constancy of the speed of light (the basis of special relativity) that has been successfully tested several times, there is a paradox between the constancy of the speed of light and other experiments that refer to spooky action at distance [13]. These experiments involve a concept called quantum entanglement. This concept simply means that when considering a pair of ("entangled") particles, when something happens to one particle, something also happens to the other particle that it is inextricably linked to, no matter how far apart the two particles are from each other (scene 7). Earlier in the last century, Einstein and other renowned physicists concluded that quantum entanglement was either impossible or that there was some hidden variable affecting the states of both particles. Einstein famously refused to believe that this phenomenon was real [14] as it violated the intricate workings of his special relativity, which implies that nothing can travel faster than the speed of light. In fact, it took quite some time for quantum entanglement to gain acceptance [15]. The detection of this phenomenon has proven to be extremely elusive, and linking particles together does appear to be impossible, but it can be achieved. In one experiment, two particles were separated by a distance of 89 miles, and the "spooky action" was still found to occur. After many other experiments, the loopholes surrounding the validity of the phenomenon were deemed closed [16, 17]. Einstein turned out to be incorrect, quantum entanglement was accepted as an actual phenomenon, and further, its real-world applications could be quite beneficial. Many physicists have continually attempted to discover methods that prove the invalidity or incompleteness of quantum entanglement; it is considered that this kind of thinking is incompatible with the previous results of experiments related to "spooky action," which may further delay the development of the applications of quantum entanglement. Meanwhile, it has become necessary to reconsider our knowledge about the nature of light; the verified existence of the "spooky action" between particles may be the end of just one chapter in the long history of physics. However, it also marks the beginning of another chapter in an exciting new field, in which the effects between two particles can be transferred faster than the speed of light. In this context, in this paper, we introduce a new mechanism that can simply explain this paradox between special relativity and spooky action (scene 8).

This paper revisits the ether problem of the Michelson–Morley experiment and provides an alternative explanation for the null result to those provided by Lorentz and Einstein [18]. The alternative solution considers that the position of the body at a given point does not necessarily imply that the body's effect must initiate from this point. In other words, there is a gap between the body's position and time, and this gap will compensate for any alteration in the speed of light created by the ether wind. Accordingly, for example, a bulb's effects (light) will not emanate from the source; instead, light is emitted at a distance away from the light source (scene 9). This new explanation can yield a better understanding of the reasons underlying quantum entanglement despite the constancy of the speed of light. Further, the explanation is consistent with the null result of the Michelson–Morley experiments, with classical mechanics, with experiments that prove the constancy of the speed of light, yields results that unify the particle world with the macroscopic world, and finally again posits the existence of ether. There are many new avenues of research can open up based on this new explanation.

The second contribution of this paper is related to simultaneity. According to Einstein, simultaneity between two events separated in space is relative; it depends on the observer's frame of reference. In contrast, in this study, simultaneity is absolute regardless of the observer's frame of reference.

The third contribution is related to time. We find that the time of the observed body progresses and regresses according to the position of the observer.

This paper is organized as follows: Sections 2 and 3 describe the Einstein train problem and the concept underlying our explanation of the null result. Sections 4 onward describe and discuss the results based on the concept.

2. Ether Problem in the Train Experiment

The null result of the Michelson–Morley experiment indicates there is a problem with regard to the existence of ether. This problem can be resolved by considering the train experiment, which is a thought experiment [19] designed by Einstein [20]. Let us first consider the "Einstein train," which moves with a velocity v, and let us assume that a bulb is located in the middle of the train's carriage at point (A). A ray of light is emitted from the middle of the carriage to a passenger at the back ("rear" passenger) located at point (B) and to a passenger at the front ("front" passenger) at point (C) of the vehicle. Let the length between (A) and (B), r_{AB}, be equal to the length between (A) and (C), r_{AC}, and let the train move in the direction of the ray emitted from (A) to (B). Further, we assume the existence of the ether, and therefore, the velocity of the train will generate an ether wind blowing opposite to the train's direction (toward point (C)) with a velocity equal to that of the train, which influences the speed of light (scene 10). We can explain this situation in terms of the two following points.

First, in accordance with classical mechanics [21] (Galilean velocity addition law), the speed of light is not constant, and additive and subtractive mathematical operations can be applied to it [22]. Therefore, the velocity of the ether wind is added to the speed of light if light moves in the same direction as the ether wind, whereas it is subtracted if light moves in the direction opposite to the ether wind. Therefore, the time required by light to travel from (A) to (C) in the latter case is given by

$$\left(t_C - t_A\right) = \frac{r_{AC}}{c + v}, \quad (1)$$

where $\left(t_A\right)$ represents the time at which the light is emitted from the bulb toward the front passenger and $\left(t_C\right)$ is the time at which the light reaches the front passenger.

Second, the time required by light to travel from (A) to (B) when it moves against the direction of the ether wind is given by

$$\left(t_B - t_A\right) = \frac{r_{AB}}{c - v},(2)$$

where $\left(t_A\right)$ represents the time at which light is emitted from the bulb and $\left(t_B\right)$ represents the time at which the light emitted from the bulb reaches the rear passenger. Thus, the two events are out of synchronicity with respect to the two moving passengers (Figure 1).

$$\left(t_B - t_A\right) \neq \left(t_C - t_A\right),(3)$$

The expected results of this experiment confirm that the light will not reach the "front" and "rear" passengers at the same time (i.e., light will not reach at the same clock's reading) (Figure 2) because of the ether wind's effect (scene 11), and this is what Michelson and Morley expected while performing their experiment. However, the result was negative, which implies there is no effect of the expected ether wind on light motion. Therefore, applying the null result of the Michelson–Morley experiment to this thought experiment would lead to a situation where the two light rays reach the "rear" and "front" passengers simultaneously (Figure 3) (scene 12).

Therefore, this problem raises the following questions: what is the mechanism that considers both the effect of the ether wind and the null result of the Michelson–Morley experiment? Does this mechanism exist? If it exists, what are its consequences? Can we test these consequences experimentally? To obtain clear answers to these questions, we attempted to develop a system that can conserve classical mechanics, solve the ether problem, and subsequently explain the Michelson–Morley experiment based on the effects predicted using this concept.

3. Space and time

Time's arrow is the direction of time in which the bodies, particles, and light are moving (scene 13). To solve the ether problem and explain the null result of the Michelson–Morley experiment, we suggest a concept that states, "time's arrow is moving along a direction at a speed equal to the speed of light." The motion of time's arrow along this direction means that the time of moving bodies and particles (moving to the future) will increase by an additional amount equal to the time that the arrow travels (scene 14), according to the following equation.

$$T_{t.a} = \left(T_{m.b}\right)_{ad},(4)$$

where $\left(T_{t.a}\right)$ is the time taken by time's arrow, while $\left(T_{m.b}\right)_{ad}$ is the increase in the time of the moving body, i.e., the additional amount of time required by the moving body to travel toward the future (scene 15).

Based on this concept, we find that the "space and time of moving bodies are separated from each other." The immediate connection between the space and time of a moving body is missing; hence, there is a gap between the space and time of a moving body (like the bulb of the train). We called this gap an "S.T.G." We can determine the length of the S.T.G of the moving body $\left(L_{gap}\right)$, i.e., the distance between space and time, using the following:

$$\left(T_{m.b}\right)_{ad} = \frac{L_{gap}}{v}, (5)$$

Using (Eq. 4), we obtain

$$T_{t.a} = \frac{L_{gap}}{v}.$$

Thus,

$$L_{gap} = T_{t.a}.v.$$

In addition, as

$$T_{t.a} = \frac{L_{t,a}}{c},$$

we obtain

$$L_{gap} = \frac{L_{t,a}v}{c}, (6)$$

where (c) is the speed of light, (v) is the velocity of the moving body, and $\left(L_{t.a}\right)$ represents the distance traveled by time's arrow equal to the length of space (L). We can use the following equation to determine the length of the S.T.G.

$$L_{gap} = \frac{Lv}{c}, (7)$$

where (L) is the length of space traveled by time's arrow.

4. Results and Discussion

4.1 Effect of ether on time

Applying the aforementioned concept to the train experiment, we find that the light reaches the "rear" and "front" passengers simultaneously. To clarify this, we consider a bulb inside a moving train. Owing to the motion of time's arrow, the time of the train is increased to forward by accessory time; hence, the space and time of the train are separated. The same applies to the bulb of the train. Owing to the motion of time's arrow, the space and time of the bulb exist separately. As a result, the impact of the bulb (represented by light) located at a point (A) in the train will not initiate from its position. Rather, it will initiate from a point located at a distance equal to the length of the S.T.G of the bulb $\left(L_{gap} \right)$ (Figure 4). We can calculate the distance at which the light travels perpendicularly $\left(d_{light} \right)$ (scene 16) using the following:

$$d_{light} = \sqrt{(L)^2 + \left(L_{gap} \right)^2} , (8)$$

With respect to the moving observer such as the rear and front passengers, the time of the bulb exceeds its position. There is always a gap between the time and position of a moving body owing to the motion of time's arrow. Hence, the result from the perspective of the moving observer will be a bulb and light beams emitted from a point located slightly in front of the bulb along the direction of the train's motion (scene 17).

Therefore, the time required by light to travel from point (A) to point (B) (Figure 5) in relation to the ether (the absolute rest) is given by

$$\left(t_B - t_A \right) = \frac{r_{AB} - L_{gap}}{c - v} . (9)$$

From (Eq. 7), we find

$$\left(t_B - t_A \right) = \frac{r_{AB} - \dfrac{r_{AB} v}{c}}{c - v} ,$$

$$\left(t_B - t_A \right) = \frac{r_{AB} \left(1 - \dfrac{v}{c} \right)}{c - v} ,$$

$$\left(t_B - t_A \right) = \frac{r_{AB} \left(\dfrac{c - v}{c} \right)}{c - v} ,$$

$$\left(t_B - t_A \right) = r_{AB} \cdot \frac{c - v}{c} \cdot \frac{1}{c - v} ,$$

$$\left(t_B - t_A\right) = \frac{r_{AB}}{c}, (10)$$

where the length of the S.T.G of the bulb is subtracted from the distance that the light travels, as the direction of the emitted light is the same as the direction of the moving train. However, the velocity of the ether is subtracted from the speed of light, as the direction of the emitted light is opposite to the direction of the ether wind.

Moreover, the time required by light to travel from point (A) to point (C) (Figure 6) in the presence of ether is given by

$$\left(t_C - t_A\right) = \frac{r_{AC} + L_{gap}}{c + v}. (11)$$

From (Eq. 7), we find

$$\left(t_C - t_A\right) = \frac{r_{AC} + \dfrac{r_{AC}v}{c}}{c + v},$$

$$\left(t_C - t_A\right) = \frac{r_{AC}\left(1 + \dfrac{v}{c}\right)}{c + v},$$

$$\left(t_C - t_A\right) = \frac{r_{AC}\left(\dfrac{c + v}{c}\right)}{c + v},$$

$$\left(t_C - t_A\right) = r_{AC} \cdot \frac{c + v}{c} \cdot \frac{1}{c + v},$$

$$\left(t_C - t_A\right) = \frac{r_{AC}}{c}, (12)$$

where the length of the S.T.G of the bulb is added to the distance the light travels, as the direction of the emitted light is opposite to the direction of the moving train. Further, the velocity of the ether is added to the speed of light, as the direction of the emitted light is the same as the direction of the ether wind.

We find that light reaches the "rear" and "front" passengers simultaneously with respect to the absolute medium of space (scene 18).

4.2 Calculation of gap time $\left(T_{gap}\right)$

We determine the time required by light to travel from the position of impact of the moving body to the position of existence of the moving body $\left(T_{gap}\right)$ in the presence of ether. We use the train experiment to show

how $\left(T_{gap}\right)$ is determined. There are two passengers in the train experiment: the rear passenger is located at point (B) and the front passenger located at point (C). The two passengers observe the same bulb located in between them at point (A). There are two rays emitted from the bulb toward the two observers. Therefore, to calculate the time required by light to travel from the bulb to each observer accurately, we need to determine the time required by light to travel through the S.T.G of the bulb $\left(T_{gap}\right)$. We find that $\left(T_{gap}\right)$ of the bulb with respect to the rear passenger differs from that $\left(T_{gap}\right)$ with respect to the front passenger owing to the presence of ether wind; therefore, the time required by light to travel along the length of the S.T.G of the bulb $\left(L_{gap}\right)$ against the ether wind (toward the rear passenger) $\left(T_{gap}\right)_{A.E}$ is given by

$$\left(T_{gap}\right)_{A.E} = \frac{L_{gap}}{c-v}, (13)$$

$$\left(T_{gap}\right)_{A.E} = \frac{\left(\dfrac{Lv}{c}\right)}{c-v},$$

$$\left(T_{gap}\right)_{A.E} = \frac{Lv}{c} \cdot \frac{1}{c-v},$$

$$\left(T_{gap}\right)_{A.E} = \frac{L}{c} \cdot \frac{v}{(c-v)},$$

$$\left(T_{gap}\right)_{A.E} = \frac{L}{c} \cdot \frac{1}{\left(\dfrac{c}{v}-1\right)}, (14)$$

The time required by light to travel along the distance that the moving body travels along the length of the S.T.G of the bulb in the direction of the ether wind (toward the front passenger) $\left(T_{gap}\right)_{W.E}$ is given by

$$\left(T_{gap}\right)_{W.E} = \frac{L_{gap}}{c+v}, (15)$$

$$\left(T_{gap}\right)_{W.E} = \frac{\left(\dfrac{Lv}{c}\right)}{c+v},$$

$$\left(T_{gap}\right)_{W.E} = \frac{Lv}{c} \cdot \frac{1}{c+v},$$

$$\left(T_{gap}\right)_{W.E} = \frac{L}{c}\cdot\frac{v}{\left(c+v\right)},$$

$$\left(T_{gap}\right)_{W.E} = \frac{L}{c}\cdot\frac{1}{\left(\dfrac{c}{v}+1\right)}, (16)$$

As shown by Eqs. 14 and 16, the value of $\left(T_{gap}\right)$ depends on the direction of ether motion (scene 19).

4.3 Advancement and delay in the time of the moving body

The existence of the S.T.G of the moving body creates a difference in time $\left(\Delta t_{p\rightarrow i}\right)$ between the position of the moving body and its impact on the surrounding space equal to $\left(T_{gap}\right)$,

$$\left(\Delta T_{p\rightarrow i}\right) = T_{gap}, (17)$$

Based on observation by humans, there is no time difference between a person's position and the time of his impact on the surrounding space. For example, if my clock shows the time as 11 pm, I can impact the surrounding space at the same time. However, this is impossible, owing to the S.T.G. Any moving body cannot impact the surrounding space at the same moment of its time. There is always a time difference between the position of the moving body and its impact. This difference refers to delay or advance in the impact time of the moving body equal to $\left(T_{gap}\right)$. Therefore, when the impact of the body is in the direction of motion, the impact time of that body will be advanced $\left(T_{a}\right)$ by a value equal to the time required by light to travel against the ether along the S.T.G $\left(T_{gap}\right)_{A.E}$ according to the following:

$$T_{a} = \left(T_{gap}\right)_{A.E}, (18)$$

Using (Eq. 13), we find

$$T_{a} = \frac{L_{gap}}{c-v}.(19)$$

However, if the impact of the moving body is against the direction of motion, the impact time of that body will be delayed $\left(T_{d}\right)$ by a value equal to the time required by light to travel with the ether along the S.T.G, according to the following equation

$$T_{d} = \left(T_{gap}\right)_{W.E}.(20)$$

Using (Eq. 15), we get

$$T_d = \frac{L_{gap}}{c+v}.(21)$$

Therefore, to determine the time required by light to travel between any two moving bodies (t_{total}) accurately, we must consider the advance and delay in the impact time of the moving body and the time required by light to travel between the positions of the moving bodies (T_l). Accordingly, the total time (t_{total}) required by light to travel from a moving body (bulb) against the direction of the ether's motion to another moving body (the rear passenger) is given by

$$t_{total} = T_l - T_a.(22)$$

From (Eq. 19), we get

$$t_{total} = \frac{L}{c-v} - \frac{L_{gap}}{c-v}.$$

From section (4.1), we get

$$t_{total} = \frac{L}{c}.(23)$$

While the total time (t_{total}) required by light to travel from the first moving body (bulb) in the direction of the ether's motion to another moving body (the front passenger) is given by

$$t_{total} = T_l + T_d,(24)$$

From (Eq. 21), we have

$$t_{total} = \frac{L}{c+v} + \frac{L_{gap}}{c+v},$$

and from section (4.1), we get

$$t_{total} = \frac{L}{c}.(25)$$

Therefore, the rays of light will simultaneously reach any two moving bodies despite the existence of the ether (scene 20).

As an example to clarify the advance and delay in the impact time of moving bodies, let us imagine a tennis match between two players. The first player wears red (hereafter, "the red player"), while the other wears blue (hereafter, "the blue player"). The match is played in the presence of ether wind. Let us imagine that the ball is a ball of light. When the red player hits the ball, it moves toward the blue player in the direction of ether wind; hence, it moves faster than expected. On the other hand, the blue player hits the ball toward the red player against the ether wind; hence, it moves slower than expected. Suppose the red player delays hitting the ball, i.e., he hits the ball at 11:30:08 pm instead of at 11:30:05 pm; this time difference (+3 s) is enough to compensate for the effect of the ether wind on the ball; hence, the ball reaches the other player depending only on its speed. Further, when the blue player hits the ball early, he hits the ball at 11:30:02 pm instead of at 11:30:05 pm and this time difference (-3 s) is enough compensate for the effect of the ether wind on the ball; hence, the ball reaches the other player depending on its speed only. Therefore, the time required by the ball to travel from the red player to the blue player or vice versa is the same (scene 21). Owing to this mechanism, the effect of ether wind on light is immeasurable.

4.4 Confusion of time of moving body

In the train scenario, let us imagine that many passengers observe the bulb. These passengers are at varying distances from the bulb. Hence, the length of the S.T.G of the bulb varies for each passenger as a result of the differing distance (L) between each observer and the bulb (Eq. 7); thus, the beginning of the bulb's impact will be delayed and advanced by different values (scene 22). Accordingly, the impact time of each moving body will be delayed and advanced by several values depending on $\left(L_{gap} \right)$ of the moving body. As $\left(L_{gap} \right)$ refers to the missing connection between space and time, time cannot be determined along this distance; therefore, the impact time of the moving body will be delayed and advanced by several values simultaneously (scene 23).

To determine the possible changes of the impact time (delays and advances) of the moving body, we need to determine the largest delay in the impact time of the moving body, i.e., the time required by light to travel against the direction of the moving body and along the S.T.G. This value is called "the maximum delay in impact time," expressed as $\left(T_d \right)$. In addition, we need to determine the largest advance in the impact time of the moving body, i.e., the time required by light to travel in the direction of the moving body and along the S.T.G. This value is called "the maximum advance in impact time," expressed as $\left(T_a \right)$. Thus, in relation to the impact of the moving body, all possible delays $\left(t_d \right)_p$ and advances $\left(t_a \right)_p$ that are equal or less than $\left(T_d \right)$ or $\left(T_a \right)$ will share the same position of the moving body (x). For example, the impact time of the train is delayed and advanced by many different values at a single position of the train. The possible advance in time of the impact of the train $\left(t_a\right)_p$ can be given by

$$\left(t_a\right)_p \leq T_a . (26)$$

From (Eq. 19), we find

$$\left(t_a\right)_p \leq \frac{L_{gap}}{c-v}.$$

From (Eq. 14), we obtain

$$\left(t_a\right)_p \leq \frac{L}{c} \cdot \frac{1}{\left(\dfrac{c}{v}-1\right)} . (27)$$

The possible delays in time of the impact of the train $\left(t_d\right)_p$ can be determined by

$$\left(t_d\right)_p \leq T_d . (28)$$

From (Eq. 21), we find

$$\left(t_d\right)_p \leq \frac{L_{gap}}{c+v}.$$

From (Eq. 16), we obtain

$$\left(t_d\right)_p \leq \frac{L}{c} \cdot \frac{1}{\left(\dfrac{c}{v}+1\right)} . (29)$$

We find that the length of the S.T.G of the moving body is directly proportional to the possible change in the impact time of that body; therefore, "the greater the length of the S.T.G, the more possible delays and advances in the impact time." In addition, the velocity of the moving body is directly proportional to the possible changes in the impact time of the moving body; hence, "the greater the velocity of the moving body, the more possible delays and advances in the impact time."

4.5 Confusion of position of the moving body

Let us apply the explanation in the previous section on possible positions of the moving body to the train scenario. Let us imagine that there are many passengers that observe the bulb, and they are located at varying distances from the bulb. In this case, the time that the light requires to travel along the

S.T.G$\left(T_{gap}\right)$ varies for each passenger as a result of the differing times (t) between each observer and the bulb, according to the following equations.

From (Eq. 7),

$$t = \frac{L_{gap}}{v}.(30)$$

Thus,

$$t = \frac{T_{gap}\left(c \pm v\right)}{v},$$

$$T_{gap} = \frac{v\,t}{c \pm v}.(31)$$

Thus, the position of the bulb will be displaced by several values along the direction of motion. Hence, the distance between the bulb and passengers is increased and decreased by several lengths. Accordingly, any moving body will be displaced by several values depending on the time required by light to travel along the S.T.G$\left(T_{gap}\right)$. As $\left(T_{gap}\right)$ refers to the missing connection between space and time, space does not exist along this distance; thus, the position of a body cannot be determined. Therefore, the moving body will be displaced by several values simultaneously.

To determine the possible displacement of the moving frame, we need to determine the largest displacement in the position of the moving body, i.e., the distance that the moving body is displaced along the length of the S.T.G $\left(L_{gap}\right)$. This value is called "the maximum displacement in position," expressed as$\left(L_{dis}\right)$, where

$$L_{dis} = L_{gap}.(32)$$

Thus, in relation to the displacement of the moving frame, all the possible lengths of displacement $\left(l_{dis}\right)_{p}$ that are equal to or less than $\left(L_{dis}\right)$ will exist at the time of the moving body (t). For example, at the time of the train, it is displaced by many different lengths simultaneously; we can determine the possible lengths that the train is displaced by $\left(l_{dis}\right)_{p}$ using the following:

$$\left(l_{dis}\right)_{p} \leq L_{dis}.(33)$$

From (Eq. 7), we find

$$\left(l_{dis}\right)_p \le \frac{Lv}{c} . (34)$$

We find that the time required by light to travel along the S.T.G is directly proportional to the possible changes in the position of the moving body; therefore, "the greater the travel time through the S.T.G $\left(T_{gap}\right)$, the more possible displacements in the position of the moving body." In addition, the velocity of the moving body is directly proportional to the possible changes in the position of the moving frame, i.e., "the greater the velocity of the moving body, the more possible displacements in the position of the body."

4.6 Act of measurement

The presence of light plays an important role in eliminating the confusion related to the position and time of the moving bodies. Light acts as an organizer that connects the possible positions and times of the moving body with the surrounding space depending on the distances and times between the moving body and the surrounding space. Therefore, (d) and (t) are equal to certain values. Hence, the possible positions and times of the moving body are actually a single position and time related to the surrounding space and time. To understand this, let us consider the analogy of a group of friends who go to a restaurant. A waiter takes the orders, and each person orders a different food item. Subsequently, the waiter arrives with a different dish for each person, the friends eat, and then leave. Let us further assume that they usually visit the restaurant every weekend. However, on one of the weekends, they do not find the waiter. The restaurant manager says, "The waiter is absent today, and so we have an open buffet system." Accordingly, the friends visit the buffet, and each one takes many portions of their favorite food. Now, it is clear that the absence of the waiter has resulted in a change in the amount of food consumed, i.e., from one portion to several portions (scene 24). Our concept can be considered analogous to the system in this restaurant. Light plays the role of the waiter, the surrounding space plays the role of the visitors, and the position and time of the moving body play the role of the food. From the perspective of the surrounding space (the observers), the possible time by which the observed body is delayed and advanced and the possible values by which the observed body is displaced will be distributed to the surrounding space (the surrounding observers) with the help of light depending on the distances and times that the light travels; thus, each point of the surrounding space will be accompanied by a single possible value of displacement of the position of the moving body, while each second of the surrounding time will be accompanied by a single possible value of a delay or advance in the time of the moving body; therefore, in the absence of light, the possible times by which the observed body is delayed or advanced and the possible values by which the observed body is displaced cannot be disturbed. Hence, confusion of the position and time of the moving frame is expected. Accordingly, we can state that the measurement process influences the obtained results. The second role that the light plays is that of a compensator. Owing to the generation of ether wind, the speed of light increases and decreases by a value equal to the velocity of the ether wind; hence, the light travels faster or slower than the speed of light (c). Hence, the light will take more or less time to travel through space. This difference in the time of travel is enough to compensate for all the delays and advances in the time of the moving body. Hence, the speed of light appears constant regardless of the motion of the light source or the presence of ether wind (scene 25). The possible change in time of the moving body can be considered as a fifth dimension, with the three dimensions of space and the time dimension, while the possible changes in the position of the moving body can be considered as a sixth dimension. The fifth and the sixth dimensions are required to describe the space and time of the moving bodies correctly.

4.7 Fifth and sixth dimensions

In the train scenario, let us imagine that the bulb is the sun, and the surrounding passengers are planets. Imagine that the speed of light is increased and decreased according to the velocity of the moving source. Hence, to calculate $\left(L_{gap}\right)$ and $\left(T_{gap}\right)$ of the sun in relation to the earth, we follow the following process.

First, in the absence of light:

   a- From the perspective of the fifth dimension, the length of the S.T.G of the sun in relation to the earth is equal to

$$\left(L_{gap}\right)_{sun} = \frac{Lv_s}{c}, (35)$$

where (L) is the length of space between the sun and earth that the arrow of time travels, while $\left(v_s\right)$ is the velocity of the sun. We find the possible advances or delays of the sun related to earth are given by

$$\left[\left(t_a\right)_p\right]_{sun} \leq \frac{L}{c} \cdot \frac{1}{\left(\dfrac{c}{v_s} - 1\right)}, (36)$$

$$\left[\left(t_d\right)_p\right]_{sun} \leq \frac{L}{c} \cdot \frac{1}{\left(\dfrac{c}{v_s} + 1\right)}. (37)$$

   b- From the perspective of the sixth dimension, the time required by light to travel along the S.T.G of the sun in relation to the earth is equal to

$$\left(T_{gap}\right)_{sun} = \frac{v_s t}{c \pm v_s}, (38)$$

where (t) is the time taken by time's arrow to travel from the sun to earth. We find the possible length that the sun is displaced in relation to the earth is given by

$$\left[\left(l_{dis}\right)_p\right]_{sun} \le \frac{Lv_s}{c}. (39)$$

Second, in the presence of light:

a- From the perspective of the fifth dimension, when the length of the S.T.G of the sun in relation to the earth is equal to

$$\left(L_{gap}\right)_{sun} = \frac{Lv_s}{c}. (40)$$

The possible times that the time of the sun is advanced are given by

$$\left[\left(t_a\right)_p\right]_{sun} = \frac{L}{c} \cdot \frac{1}{\left(\dfrac{c}{v_s} - 1\right)}. (41)$$

Using (Eq. 22), the total time required by light to travel along the length of space between the sun and earth will be given by

$$t_{total} = \frac{L}{c - v_s} - \left[\left(t_a\right)_p\right]_{sun}, (42)$$

$$t_{total} = \frac{L}{c}. (43)$$

The possible times that the time of the sun is delayed are given by

$$\left[\left(t_d\right)_p\right]_{sun} = \frac{L}{c} \cdot \frac{1}{\left(\dfrac{c}{v_s} + 1\right)}. (44)$$

Using (Eq. 24), the total time required by light to travel along the length of space between the sun and earth will be given by

$$t_{total} = \frac{L}{c + v_s} + \left[\left(t_d\right)_p\right]_{sun}, (45)$$

$$t_{total} = \frac{L}{c}. (46)$$

b- From the perspective of the sixth dimension, the time required by light to travel along the S.T.G of the sun in relation to the earth is equal to

$$\left(T_{gap}\right)_{sun} = \frac{v_s\, t}{c \pm v_s}\,.(47)$$

The possible length that the position of the sun is displaced in relation to the earth is given by

$$\left[\left(l_{dis}\right)_p\right]_{sun} = \frac{L v_s}{c}\,.(48)$$

Therefore, the total length that light travels along the length of space between the sun and earth is given by

$$L_{total} = L - \left[\left(l_{dis}\right)_p\right]_{sun}\,,(49)$$

$$L_{total} = L + \left[\left(l_{dis}\right)_p\right]_{sun}\,,(50)$$

where

$$\left(l_{dis}\right)_p \cdot \left(t_a\right)_p = c - v\,,(51)$$

$$\left(l_{dis}\right)_p \cdot \left(t_d\right)_p = c + v\,.(52)$$

Using the equations in this section, the various positions of the sun provide a single unified time required by light to travel from the sun to earth regardless of the possible positions of the sun relative to earth (scene 26).

4.8 Entanglement

According to the previous section (4.7), we set the S.T.G of the sun relative to the earth as $\left(L_{gap}\right)_{sun}$. Now, let us imagine there is a moving body is directed toward the sun where the distance (d) between them becomes equal to or less than $\left(L_{gap}\right)_{sun}$ or the time (t) between them is equal to or less than $\left(T_{gap}\right)_{sun}$,

$$d \leq \left(L_{gap}\right)_{sun}\,,(53)$$

$$t \leq \left(T_{gap}\right)_{sun}\,,(54)$$

In this case the expected result is the entanglement of the two bodies related to the earth. The connection between them does not require time related to the earth; the two observed bodies affect each other without the need for light to transfer data between them. We explain quantum entanglement based on this concept (scene 27).

4.9 Nature of time

We can discover the nature of time and answer the following questions: Does relativity of time exist? How we can determine time according to the observer's frame of reference? All these questions need certain and clear answers.

We classify the frames of reference into three categories: moving frame of reference, absolutely stationary frame of reference, and relatively stationary frame of reference.

4.9.1 Moving frame of reference

The time of the moving frame with respect to the moving observer (t_m) (observer in a moving frame that is stationary with respect to itself; for example, the passengers on a train) can be determined according to the length of the S.T.G (L_{gap}), length of space (L), and speed of light (c) (scene 28), as per the following equation.

$$\left(ct_m\right)^2 = \left(L\right)^2 + \left(L_{gap}\right)^2 ,(55)$$

$$t_m{}^2 = \frac{L^2 + L_{gap}{}^2}{c^2},$$

$$t_m{}^2 = \frac{L^2 + \left(\dfrac{Lv}{c}\right)^2}{c^2},$$

$$t_m{}^2 = \frac{1}{c^2}\left(L^2 + \left(\frac{Lv}{c}\right)^2\right),$$

$$t_m{}^2 = \frac{1}{c^2}\left(L^2 + \frac{L^2 v^2}{c^2}\right),$$

$$t_m{}^2 = \frac{1}{c^2}\left(L^2\left(1 + \frac{v^2}{c^2}\right)\right),$$

$$t_m{}^2 = \frac{L^2}{c^2}\left(1 + \frac{v^2}{c^2}\right),$$

$$t_m = \frac{L}{c}\sqrt{1 + \frac{v^2}{c^2}}.(56)$$

Therefore, the time of the moving observer is dilated related to itself by an amount equal to $\sqrt{1+\dfrac{v^2}{c^2}}$.

4.9.2 Absolutely stationary frame of reference (Ether)

The time of the moving frame with respect to the ether medium (t_e) can be determined depending on time of the moving frame with respect to the moving observer (t_m) (scene 29) using the following equations.

$$\left(c\,t_e\right)^2 = \left(c\,t_m\right)^2 + \left(v\,t_e\right)^2 , (57)$$

$$\left(c\,t_m\right)^2 = \left(c\,t_e\right)^2 - \left(v\,t_e\right)^2 ,$$

$$c^2\,\frac{t_m^{\,2}}{t_e^{\,2}} = c^2 - v^2 ,$$

$$\frac{t_m^{\,2}}{t_e^{\,2}} = \frac{c^2 - v^2}{c^2} ,$$

$$\frac{t_m^{\,2}}{t_e^{\,2}} = 1 - \frac{v^2}{c^2} ,$$

$$\frac{t_m}{t_e} = \sqrt{1 - \frac{v^2}{c^2}} ,$$

Thus,

$$t_e = \frac{t_m}{\sqrt{1 - \dfrac{v^2}{c^2}}} . (58)$$

Using this equation, we can determine that the time of the moving frame is more dilated with respect to that of the ether. We can determine (t_e) directly using the following equation.

$$t_e = \frac{l}{c} \cdot \frac{\sqrt{1 + \dfrac{v^2}{c^2}}}{\sqrt{1 - \dfrac{v^2}{c^2}}} . (59)$$

4.9.3 Relatively stationary frame of reference

The time of the moving frame with respect to the stationary observer (t_s) (who is in a moving frame that is stationary relative to another moving frame; for example, an observer on a platform observing a moving train) can be determined according to $\left(t_m\right)$, the distance that light travels through space $\left(L\right)$, the velocity of the moving frame, and the speed of light $\left(c\right)$ (scene 30), according to the following equations.

$$\left(c\,t_s\right)^2 = \left(c\,t_m\right)^2 + \left(v\,t_s\right)^2 ,(60)$$

$$\left(c\,t_m\right)^2 = \left(c\,t_s\right)^2 - \left(v\,t_s\right)^2 ,$$

$$c^2 \frac{t_m^{\,2}}{t_s^{\,2}} = c^2 - v^2 ,$$

$$\frac{t_m^{\,2}}{t_s^{\,2}} = \frac{c^2 - v^2}{c^2} ,$$

$$\frac{t_m^{\,2}}{t_s^{\,2}} = 1 - \frac{v^2}{c^2} ,$$

$$\frac{t_m}{t_s} = \sqrt{1 - \frac{v^2}{c^2}} ,$$

Thus,

$$t_s = \frac{t_m}{\sqrt{1 - \dfrac{v^2}{c^2}}}.(61)$$

From this equation, we find that the time of the moving frame is more dilated with respect to the relatively stationary observer (for example, the second in the train is more dilated with respect to the observer on the station's platform than that with respect to the passengers of the train).

4.10 Nature of simultaneity

We find that simultaneity is absolute regardless of the observer's frame of reference. In Section 4.9, we categorized the frames of reference into three types, and each category has corresponding equations that determine the nature of simultaneity. Here, we consider the mathematical aspect of simultaneity as below:

4.10.1 Ether's reference

With respect to the ether, the time required by light to travel longitudinally along the direction of ether in a moving frame, can be given by

$$\left(c + v\right)t_e = L + L_{gap} ,\left(62\right)$$

$$t_e = \frac{L + L_{gap}}{c + v},$$

$$t_e = \frac{L}{c}.\,(63)$$

Meanwhile, the time required by light to travel longitudinally against the ether is given by

$$(c - v)t_e = L - L_{gap},\,(64)$$

$$t_e = \frac{L - L_{gap}}{c - v},$$

$$t_e = \frac{L}{c}.\,(65)$$

4.10.2 Moving frame of reference

With respect to the moving observer, the time required by light to travel longitudinally along the ether direction in a moving frame is given by

$$(c + v)t_m = \left(L + L_{gap}\right),\,(66)$$

$$t_m = \frac{L + L_{gap}}{c + v},$$

$$t_m = \frac{L}{c},\,(67)$$

and the time required by light to travel longitudinally against the ether direction is given by

$$(c - v)t_m = \left(L - L_{gap}\right),\,(68)$$

$$t_m = \frac{L - L_{gap}}{c - v},$$

$$t_m = \frac{L}{c}.\,(69)$$

4.10.3 Relatively stationary frame of reference

With respect to the stationary observer, the time required by light to travel longitudinally along the ether direction in a moving frame can be expressed as

$$c\,t_s = \left(L + L_{gap}\right) - \left(v\,t_s\right),\,(70)$$

$$\left(c\,t_s\right)+\left(v\,t_s\right)=L+L_{gap},$$

$$t_s\left(c+v\right)=L+L_{gap},$$

$$t_s=\frac{L+L_{gap}}{c+v},$$

$$t_s=\frac{L}{c}.(71)$$

However, the time required by light to travel longitudinally against the ether direction is

$$c\,t_s=\left(L-L_{gap}\right)+\left(v\,t_s\right),(72)$$

$$\left(c\,t_s\right)-\left(v\,t_s\right)=L-L_{gap},$$

$$t_s\left(c-v\right)=L-L_{gap},$$

$$t_s=\frac{L-L_{gap}}{c-v},$$

$$t_s=\frac{L}{c}.(73)$$

Therefore, the simultaneity is absolute regardless the observer's frame of reference.

4.11 Negation of ether effect on light motion

The concept underlying the Michelson–Morley experiment is similar to that of the previously mentioned train experiment. As is well known, Maxwell's equation [23] can be used to derive a wave equation to predict the speed of light [24, 25], but it is unclear as to what frame this is possible in. For a sound wave, the acoustic wave equation can be used to predict the sound speed; again, what frame is being referred to here? In the frame of stationary air, it has been hypothesized that the ether exists and plays the role of the carrier of electromagnetic radiation [26]; hence, when electromagnetic waves propagate in ether, the velocity with which the waves spread depends on the ether and not on the velocity of the wave source. This means that when we measure the velocity of electromagnetic waves, the result depends on the velocity of the measurement apparatus; hence, if the speed of the waves is the same as that of the ether wind, it does not imply that the speed of the waves is the same as that of the apparatus being used to measure it [27]. Using this concept, Michelson and Morley designed an experiment to detect the ether and measure its effect on the speed of light. To achieve this, they used interferometer techniques [28], telescopes, light sources, two reflecting mirrors, and a beam splitter [29]. The experiment is initiated at the light source where the light is emitted in the direction of the beam splitter.

A portion of the light beams is reflected transversely toward the first reflecting mirror and another portion is transmitted horizontally to a reflecting mirror, but both beams merge at the beam splitter and reach the telescope together [30]. The two beams form interference fringes as a result of their phase difference. These interference fringes are recorded, and the apparatus is then rotated by 90° after which the interference is recorded again (Figure 7). A fringe shift of 0.4 is expected, but the experimentally measured fringe shift is 0.01 [31]. Here, we remark that the Michelson–Morley experiment cannot prove the existence of ether [32].

On the other hand, let us consider the mathematical viewpoint of this experiment. As the earth moves through stationary ether, an ether wind is generated in the direction opposite to the earth's motion. Hence, we can calculate the time that the first beam takes to move longitudinally to the reflecting mirror and return to the beam splitter. Michelson and Morley calculated the required time using the following equation.

$$T_l = \frac{L}{c+v} + \frac{L}{c-v}. (74)$$

They calculated the travel time of the second beam moving transversely with respect to the first beam using the following equation

$$T_t = \frac{2L}{\sqrt{c^2 - v^2}}. (75)$$

Hence, there is a time difference between the two beams that can be expressed as

$$T_l - T_t = \left(\frac{L}{c+v} + \frac{L}{c-v} \right) - \left(\frac{2L}{\sqrt{c^2 - v^2}} \right),$$

$$T_l - T_t = \left[\frac{2L}{c} \cdot \frac{1}{1 - \frac{v^2}{c^2}} \right] - \left[\frac{2L}{c} \cdot \frac{1}{\sqrt{1 - \frac{v^2}{c^2}}} \right],$$

$$T_l - T_t = \frac{2}{c} \left[\frac{L}{1 - \frac{v^2}{c^2}} \right] - \left[\frac{L}{\sqrt{1 - \frac{v^2}{c^2}}} \right]. (76)$$

However, the result of their experiment was negative. There was no time difference between the travel times of the two beams; the experimental result did not match the expected value calculated by Michelson and Morley (Eq. 76). Now, let us apply the obtained results (discussed in Section 4.1) to this experiment:

(a) The time required by light to travel longitudinally from the beam splitter to the first reflecting mirror and return (T_l) with respect to the moving frame of reference can be determined as

"Time required by light to travel longitudinally $\times$ Time dilation of the moving frame"

$$(T_m)_l = \left(\frac{L + l_{dis}}{c + v} + \frac{L - l_{dis}}{c - v} \right) \cdot \sqrt{1 + \frac{v^2}{c^2}} \cdot (77)$$

Thus,

$$(T_m)_l = \frac{2L}{c} \cdot \sqrt{1 + \frac{v^2}{c^2}} \cdot (78)$$

Accordingly, we can determine (T_l) with respect to time of the absolutely stationary frame (ether), using the following.

$$(T_e)_l = \frac{t_m}{\sqrt{1 - \dfrac{v^2}{c^2}}} \cdot (79)$$

(b) The time required by light to travel transversely from the beam splitter to the second reflecting mirror and return (T_t) with respect to the moving frame of reference can be determined using

"Time required by light to travel transversely (not affected by ether) $\times$ Time dilation of the moving frame"

$$(T_m)_t = \frac{2L}{c} \cdot \sqrt{1 + \frac{v^2}{c^2}}, (80)$$

Hence, (T_t) with respect to the time of the ether is given by

$$(T_n)_t = \frac{t_m}{\sqrt{1 - \dfrac{v^2}{c^2}}}, (81)$$

From (Eq. 79 and 81), we note that

$$T_l = T_t \cdot (82)$$

(Eq. 82) indicates that the time required by light to travel longitudinally from the beam splitter to the first reflecting mirror and return (T_l) equals the time required by light to travel transversely from the beam splitter to the second reflecting mirror and return (T_t). The distance that the moving frame travels through time (L_{ga}) always compensates for any alteration in the speed of light due to the ether wind, i.e., (L_{gao}) negates the effect of ether wind on light motion and makes it "unmeasurable" in experiments. Hence, the result of the Michelson–Morley experiment is negative despite the existence of the ether.

5. Conclusions

Based on the results of this paper, we can conclude that the delay or advance in the time of impact of moving bodies refers to progression and regression in the time of moving bodies. From the previous example in section (4.3), the red player delays to hit the ball by 3 s, this refers to the time at which the red player progresses 3 s forward from the perspective of the blue player. Hence, the impact of the red player is exerted late from the perspective of the blue player. We find that time has the ability to progress and regress depending on the distance between the moving body and the surrounding space, and due to the multiple possible advances and delays that exist simultaneously, the time of the moving body will progress and regress by different values simultaneously as

$$t_{regress} = t_{rest} - T_a , (83)$$

$$t_{progress} = t_{rest} + T_d , (84)$$

where $\left(t_{progress}\right)$ is the value of the progression in time, $\left(t_{regress}\right)$ is the value of the regression in time, and $\left(t_{rest}\right)$ is the time of the body at rest.

Therefore, the clock reading will be confused, and the time of the moving body is confused in nature; hence, "the greater the advance and delay in time relating to the surrounding space, the more the progression and regression in time relating to the surrounding space (scene 31)."

We remark that ether, light, and the S.T.G. can be considered a triple-point basis for most physical laws. We call this triple-point basis the "pyramid of physics" in the sense that we cannot obtain a complete understanding of physics without these aspects.

Acknowledgements

I wish to express my sincere gratitude to all the heroes of the Egyptian revolution on 25th January who sacrificed their lives for "bread, freedom, and social justice." They were a source of inspiration for me in terms of the importance of seeking humanity to uphold the values and principles of society as well as freedom. They are the real motivation behind the completion of this long and hard study, which took me six years to complete.

Funding

This research did not receive any specific grant from funding agencies in the public, commercial, or not-for-profit sectors.

Conflict of Interest

None

References

[1]   Michelson, A.A., Morley, E.W.: On the relative motion of the Earth and the luminiferous ether. Am. J. Sci. 34, 333–345 (1887)

[2]   Michelson, A.A., Morley, E.W.: Influence of motion of the medium on the velocity of light. Am. J. Sci., 31, 377–385 (1886)

[3]   Gorbatsevich, F.F.: The Ether and Universe. VDM Verlag, Saarbrücken (2010)

[4]   Whittaker, E.T.: A History of the Theories of Aether and Electricity from the Age of Descartes to the Close of the Nineteenth Century. Longmans, Green, and Co., London, New York, Bombay, and Calcutta (1910)

[5]   Lorentz, H.A.: The relative motion of the earth and the ether. https://en.wikisource.org/wiki/Translation:The_Relative_Motion_of_the_Earth_and_the_Aether (2015). Accessed July 2015

[6]   Cushing, J.T.: Vector Lorentz transformations. Am. J. Phys. 35, 858 (1967)

[7]   Lorentz, H.A.: Electromagnetic phenomena in a system moving with any velocity smaller than that of light. Proc. R. Neth. Acad. Arts Sci. 6, 809–831 (1904)

[8]   Einstein, A.: Zur Elektrodynamik bewegter Körper. Ann. Phys. 322, 891–921 (1905)

[9]   Darrigol, O.: Einstein, 1905-2005. In: Damour, T., Darrigol, O., Duplantier, B., Rivasseau, V. (eds.) Einstein, pp. 1905-2005: Poincare Seminar (2005), Birkhäuser Verlag, Basel (2006)

[10]   Einstein, A.: Relativity: The Special and General Theory. Henry Holt and Company, New York (1920)

[11]   Ryder, P.: Classical Mechanics. Shaker Verlag, GmbH, Germany (2007)

[12]   McEvoy, P.: Classical Theory (Theory of Interacting systems). MicroAnalytix (2002)

[13]   Hesse, M. B: Forces and Fields: The Concept of Action At A Distance in the History of Physics, Thomas Nelson and Sons, London (1969).

[14]   Einstein, A., Podolsky, B., Rosen, N.: Can a quantum-mechanical description of physical reality be considered complete? Phys. Rev. 47, 777–780 (1935)

[15]   Bell, J.S.: On the problem of hidden variables in quantum mechanics. Rev. Mod. Phys. 38, 447–452 (1966)

[16]   Giustina, M., et al.: A significant loophole-free test of Bell's theorem with entangled photons. Phys. Lett. (2015) (submitted).

[17]   Hensen, B., et al.: Loophole-free Bell inequality violation using electronspins separated by 1.3 kilometers. Nature 526, 682–686 (2015)

[18]   Phillips C., Priwer S.; The Everything Einstein Book. Adams Media Corporation, USA (2003)

[19]   Comstock, D.F.: The principle of relativity. Science 31, 767–772 (1910)

[20]   Rosser, W.G.V.: Introductory Special Relativity. CRC Press, Las Vegas, NV (1992)

[21]   Einstein, A.: Dialog über Einwände gegen die Relativitätstheorie. Naturwissenschaften 6, 697–702 (1918)

[22]   Stachel, J.: EINSTEIN and MICHELSON The Context of Discovery and the Context of Justification. Astron. Nachr. 303, 47–53 (1982)

[23]   Maxwell, J.C.: Encyclopædia Britannica. C. Scribner's Sons, New York (1878)

[24]   Cook, D.M.: The Theory of the Electromagnetic Field. Courier Dover Publications, New York (2002)

[25]   Simhony, M.: Invitation to the Natural Physics of Matter, Space, and Radiation. World Scientific, Singapore (1994)

[26]   Genz, H.: Nothingness: The Science of Empty Space. Basic Books, New York (2001).

[27]   Schurnacher, R.A.: Special relativity and the Michelson–Morley interferometer. Am. J. Phys. 62, 609 (1994)

[28]   Sheriff, P.: Michelson-Morley Experiment. CreateSpace Independent Publishing Platform (2015)

[29]   Janssen, M., Stachel, J.: The Optics and Electrodynamics of Moving Bodies. Max Planck Institut Für Wissenschaftsgeschichte, Germany (2004)

[30]   Essen, L.: A new Æther-drift experiment. Nature 175, 793–794 (1955)

[31]   Swenson, L.S.: The Michelson-Morley-Miller experiments before and after 1905. J. Hist. Astron. 1, 56–78 (1970)

[32]   Woodhouse, N.M.J.: Special Relativity. Springer Science & Business Media, Berlin (2007)

Figure Captions

Fig. 1 Rays of light emitted from the middle of the train reach the "front" passenger (C) before the "rear" passenger (B) as a result of the effect of the ether wind on the motion of light

Fig. 2 Asynchronicity of the "rear" and "front" clocks from the perspective of the moving observer; these readings are only used to clarify the idea of the Michelson–Morley experiment

Fig. 3 Synchronicity of the "rear" and "front" clocks from the perspective of the moving observer; these readings are used to only clarify the idea of the Michelson–Morley experiment

Fig. 4 Existence of distance between the bulb and its effect (emanation of light)

Fig. 5 Light from the bulb emitted from point (X) that is located closer to the "rear" passenger than the "front" passenger

Fig. 6 Light from the bulb emitted from point (X) located far from the "front" passenger

Fig. 7 Schematic of the Michelson–Morley experiment

YOUR KNOWLEDGE HAS VALUE

- We will publish your bachelor's and
  master's thesis, essays and papers

- Your own eBook and book -
  sold worldwide in all relevant shops

- Earn money with each sale

Upload your text at www.GRIN.com
and publish for free